First Questions and Answers about **Bedtime**

Where Does the Sun Sleep?

TIME LIFE for Children®

ALEXANDRIA, VIRGINIA

Contents

Why do I yawn when I get tired?

A yawn is one way in which your body tries to stay awake when you are sleepy. When you yawn, you take in a lot of air. That extra air keeps you from falling asleep. A yawn is also a signal to your mom or dad. It's time to get ready for bed!

4

Did you know?
People aren't the only ones who yawn. Many animals yawn, too.

Why do I have to go to sleep?

Sleep gives your body a chance to rest. After a day of running, jumping, laughing, eating, and talking, your body is tired. While you are asleep, your body can heal cuts, bruises, and sore muscles.

Little babies sleep almost all the time. Children your age don't sleep as much as babies. Grownups need less sleep than children your age.

What do you do to get ready for bed at night?

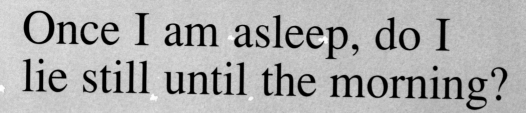

Once I am asleep, do I lie still until the morning?

Probably not. People move during the night. They turn their heads and move their arms and legs. Even though you are sleeping, your body is not completely still. You may go to sleep lying on your back, but end up on your belly.

Did you know?

Some people talk in their sleep! And sometimes people get up and walk while they are still asleep. People who talk or walk in their sleep don't even know they are doing it.

Where do animals sleep?

Birds sleep perched on tree limbs.

Animals sleep in special places, just like you. But they don't sleep in a bed the way you do. From the smallest to the biggest, animals sleep where they feel comfortable and safe.

Woodchucks and **Chipmunks** sleep in burrows in the ground.

Squirrels sleep
in huge round
nests they make
from leaves.

Turtles nap on logs.

Fish sleep near the bottom of rivers
and ponds. But they sleep with their
eyes wide open!

11

Where does a whale sleep?

Whales sleep just below the water's surface in the ocean. They move slowly in a circle while they doze. But they need to breathe air, just as you do. So, even when they're resting, they come up to the surface to breathe every once in a while.

13

Where does the sun sleep?

The sun doesn't really sleep. We just can't see it for a while. We live on a world that is shaped like a giant ball, and it turns all the time. During the day, when the sky is filled with sunlight, our side of the world faces the sun.

Good night!

As the day passes, the world turns. Now the other side faces the sun. On our side the sky is dark and filled with stars. It is nighttime. When it's nighttime for us, it's daytime for people who live on the other side of the world!

15

Do flowers and trees sleep?

In a way they do. Leaves that are stiff in the sunlight droop in the dark. Flowers such as daisies and tulips close up their petals.

Did you know?
Some flowers wake up at night. The sweet-smelling honeysuckle opens its petals when the sun goes away.

Who is awake at night?

Most people work or play during the day and sleep at night. But some people work during the night and sleep at other times of day.

Did you know?
People who like to stay awake at night are called night owls, because owls stay awake all night.

Whooo!

The **baker** works at night so fresh bread will be ready in the morning.

Some **police officers** and **firefighters** stay awake. If someone needs them during the night, they are ready to help.

Long before it gets light, **farmers** bring fresh fruit and vegetables to market.

Some restaurants are open all night.

Does everyone else sleep?

Plenty of other people stay awake at night.

In the hospital, nurses and doctors care for sick people at night.

MORPHE
BEDS

Truckdrivers travel both day and night.

Radio stations broadcast news, weather, and music to listeners all night.

Will you help me find my way back home?

City streets are cleaned at night.

Do animals stay awake at night?

Some animals stay awake at night and sleep all day.

Whooo!

Owls stay awake at night to hunt for mice and other small animals.

Many insects are busy at night. **Crickets** chirp. **Fireflies** light up. **Moths** flutter around street lamps.

Chirp!

Garden **spiders** spin their webs after dark.

Bats eat insects that come out at night.

After dark, **raccoons** search for scraps of food.

Ribbit!

CRASH!

At night, noisy **bullfrogs** croak at the water's edge.

23

How do animals see at night?

Many night animals have large eyes. This helps them see even when there is little light. Animals that come out at night also use their noses to help them find things in the dark.

Did you know?

Look at a cat's eyes at night. When light reaches them, they seem to shine. A cat's eyes are specially made to use very little light. Deer, opossums, foxes, and skunks also have eyes that shine.

Who makes all that noise at night?

Crickets and other noisy insects chirp loudly at night. They chirp to call to one another. Crickets make this noise by rubbing their wings together.

Did you know?
Crickets chirp faster when the weather is warmer.

Chirp!

Why do fireflies glow in the dark?

Can you guess why they're called fireflies?

Insects have different ways of finding one another at night. Crickets chirp. And, when it is dark out, fireflies turn on their lights. A firefly flashes the light in its belly on and off. When another firefly sees the light, it flashes back.

Why can we see stars only at night?

Stars are in the sky all the time, but we cannot see them until nighttime. During the day the sky is filled with sunlight. It is so bright that we cannot see the stars. At night, when the sun isn't shining on us, the sky darkens. Then we can see the stars.

Why doesn't the moon fall out of the sky?

The moon is far away in outer space. It circles the earth every month, following the same path each time. The moon doesn't fall to earth because it is moving too fast as it circles around. But it won't ever fly away, because the earth always pulls on it a little. The earth holds the moon in its path.

32

Why does the moon change shape?

The moon is like a round mirror. It shines only when light from the sun bounces off it. The sun shines on just one part of the moon at a time. We can see only the part of the moon that is lighted by the sun. The part of the moon that faces away from the sun doesn't shine. The rest of the moon is there, but it is too dark to see.

Because the moon moves
around the earth every month,
we see a different part of it
each day. That is why the
moon seems to change shape.

How long would it take to fly to the moon?

It takes a spaceship about three days to reach the moon. The moon is far, far away, even though it looks as if it is just over the next hill. No bird or bug or bat can fly to the moon. No airplane can reach it either.

Did you know?
Astronauts have gone to the moon
six times. They gathered moonrocks
and brought them back to earth.

Why is the moon smiling at me?

It only looks as if the moon has a big face that is smiling down at you. But it's not a face at all. The moon is not smooth. It is covered with big holes called craters. It also has tall mountains and flat, gray land. When you look at the moon from earth, the mountains and craters make shadows and shapes that can trick your eyes. They make the moon look as if it has a face.

Try it!

The next time the moon is out, take a look at it. Do you see the shadows and shapes on it? Some people say they look like a face. What do they look like to you?

Why does my room look different in the dark?

When the light is on, your room looks bright and cheery. But when you turn off the light, things seem to change. Instead of bright colors, you see dark shadows.

Our eyes need light to see bright colors and clear pictures. When there is little light, it's hard to see. Shadowy shapes can mix together. Sometimes, we are not sure what we are looking at. But even though it looks different to you, your room is the same in the day as it is at night.

41

What is a dream?

A dream is like a make-believe story. You seem to see it while you are sleeping. But it doesn't happen in real life. Almost anything can happen in a dream. You may dream that you are opening a present, eating ice cream, or riding an elephant. Sometimes dreams are fun. Once in a while they are a little scary. But you don't have to worry about a dream. It always ends when you wake up.

Do animals dream?

We don't know for sure if animals dream, because they can't tell us. But when people dream, their eyes move back and forth under their eyelids. Some animals do the same thing when they sleep. They could be dreaming. What do you think animals might dream about?

What makes me wake up?

Every night the sun sets, and every morning the sun rises again. Your body is used to the way things happen every day. It is as if you had a clock inside you. After sleeping all night, your body is rested. The morning light is a signal. It's time to get up!

TIME-LIFE for CHILDREN ®

Assistant Managing Editor: Patricia Daniels
Editorial Directors: Jean Burke Crawford, Allan Fallow,
Karin Kinney, Sara Mark, Elizabeth Ward
Publishing Assistant: Marike van der Veen
Production Manager: Marlene Zack
Senior Copyeditor: Colette Stockum
Production: Celia Beattie
Supervisor of Quality Control: James King
Assistant Supervisor of Quality Control: Miriam Newton
Library: Louise D. Forstall

Special Contributor: Barbara Klein
Researcher: Catherine G. Kunkel
Writer: Andrew Gutelle

Designed by: **David Bennett Books**

Series design: David Bennett
Book design: Andrew Crowson
Art direction: David Bennett & Andrew Crowson
Illustrated by: Andrew Ellis
Additional cover
illustrations by: Malcolm Livingstone

First printing. Printed in U.S.A.
Published simultaneously in Canada.

Time Life Inc. is a wholly owned subsidiary of THE TIME INC. BOOK COMPANY.

TIME-LIFE is a trademark of Time Warner Inc. U.S.A.

For subscription information, call 1-800-621-7026.

Library of Congress Cataloging-in-Publication Data

Where does the sun sleep? : first questions and answers about bedtime.
p. cm.— (Time-Life library of first questions and answers)
Summary : Answers questions about the world at night, including
whether people move in their sleep, whether flowers and trees sleep,
why the moon changes shape, and why fireflies glow in the dark.
ISBN 0-7835-0866-2.— ISBN 0-7835-0867-0 (lib. bdg.)
1. Sleep — Juvenile literature. 2. Night — Juvenile literature.
[1. Sleep — Miscellanea. 2. Night — Miscellanea.
3. Bedtime — Miscellanea. 4. Questions and
answers.] I Time-Life for Children (Firm) II.Series: Library of
first questions and answers.
QP37.W43 1993
612 — dc20
93-6654
CIP
AC

Consultants

Dr. Lewis P. Lipsitt, an internationally recognized specialist on childhood development, was the 1990 recipient of the Nicholas Hobbs Award for science in the service of children. He has served as the science director for the American Psychological Association and is a professor of psychology and medical science at Brown University, where he is director of the Child Study Center.

Thomas D. Mullin directs the Hidden Oaks Nature Center in Fairfax County, Virginia, where he coordinates workshops and seminars designed to promote appreciation for wildlife and the environment. He is also the Washington representative for the National Association for Interpretation, a professional organization for naturalists involved in public education.

Dr. Judith A. Schickedanz, an authority on the education of preschool children, is an associate professor of early childhood education at the Boston University School of Education, where she also directs the Early Childhood Learning Laboratory. Her published work includes *More Than the ABC's: Early Stages of Reading and Writing Development* as well as several textbooks and many scholarly papers.